This Book Belongs to

____

____

Measurement Conversions

Length
1 foot = 12 inches
1 yard = 3 feet
1 mile = 5,280 feet
1 inch = 2.54 centimeters
1 foot = 30.48 centimeters
1 meter = 3.2808 feet
1 kilometer = 1,000 meters
1 kilometer = 0.621 miles
1 mile = 1.6103 kilometers

Volume
1 pint = 2 cups
1 quart = 2 pints
1 gallon = 4 quarts

Weight
1 pound = 16 ounces
1 ton = 2,000 pounds
1 ounce = 28.35 grams
1 kilogram = 1,000 grams
1 pound = 0.4536 kilograms
1 kilogram = 2.2046 pounds

Temperature
Celsius (C) = 5/9 * (F-32)
Fahrenheit (F) = (9/5 * C) + 32

Newton's Laws of Motion

1. An object at rest tends to remain at rest, and an object in motion tends to remain in motion.

2. Force is equal to mass times acceleration.

3. Every action has an equal and opposite reaction.

Periodic Table

1 **H**								
3 **Li**	4 **Be**							
11 **Na**	12 **Mg**							
19 **K**	20 **Ca**	21 **Sc**	22 **Ti**	23 **V**	24 **Cr**	25 **Mn**	26 **Fe**	27 **Co**
37 **Rb**	38 **Sr**	39 **Y**	40 **Zr**	41 **Nb**	42 **Mo**	43 **Tc**	44 **Ru**	45 **Rh**
55 **Cs**	56 **Ba**	*	72 **Hf**	73 **Ta**	74 **W**	75 **Re**	76 **Os**	77 **Ir**
87 **Fr**	88 **Ra**	**	104 **Rf**	105 **Db**	106 **Sg**	107 **Bh**	108 **Hs**	109 **Mt**

***Lanthanides**		57 **La**	58 **Ce**	59 **Pr**	60 **Nd**	61 **Pm**	62 **Sm**
****Actinides**		89 **Ac**	90 **Th**	91 **Pa**	92 **U**	93 **Np**	94 **Po**

# of Elements

																	2 He
												5 B	6 C	7 N	8 O	9 F	10 Ne
												13 Al	14 Si	15 P	16 S	17 Cl	18 Ar
28 Ni	29 Cu	30 Zn										31 Ga	32 Ge	33 As	34 Se	35 Br	36 Kr
46 Pd	47 Ag	48 Cd										49 In	50 Sn	51 Sb	52 Te	53 I	54 Xe
78 Pt	79 Au	80 Hg										81 Tl	82 Pb	83 Bi	84 Po	85 At	86 Rn
110 Ds	111 Rg	112 Cn										113 Uut	114 Fl	115 Uup	116 Lv	117 Uus	118 Uuo

63 Eu	64 Gd	65 Tb	66 Dy	67 Ho	68 Er	69 Tm	70 Yb	71 Lu
95 Am	96 Cm	97 Bk	98 Cf	99 Es	100 Fm	101 Ma	102 No	103 Lr

The Scientific Method

Step 1. Ask a question

Step 2. Do background research

Step 3. Construct a hypothesis

Step 4. Do an experiment

Step 5. Draw a conclusion

Step 6. Communicate results

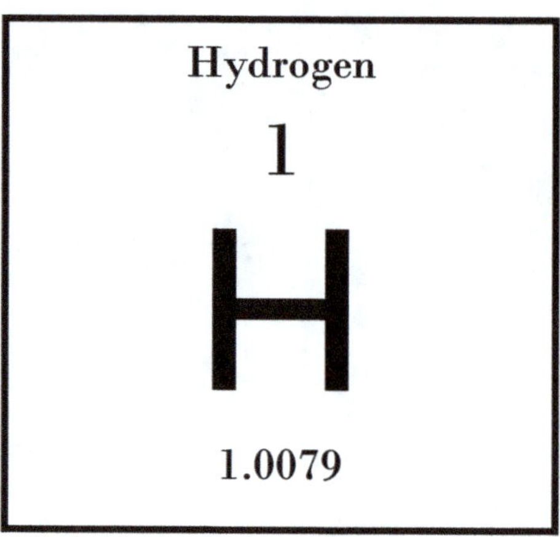

Hydrogen
1
H
1.0079

Lab Notes

Helium
2
He
4.0026

Lab Notes

Lithium
3
Li
6.941

Lab Notes

BIG IDEAS

It is the function of science to discover the existence of a general reign of order in nature and to find the causes governing this order.

— Dmitri Mendeleev

Beryllium

4

Be

9.0122

Lab Notes

Boron
5
B
10.811

Lab Notes

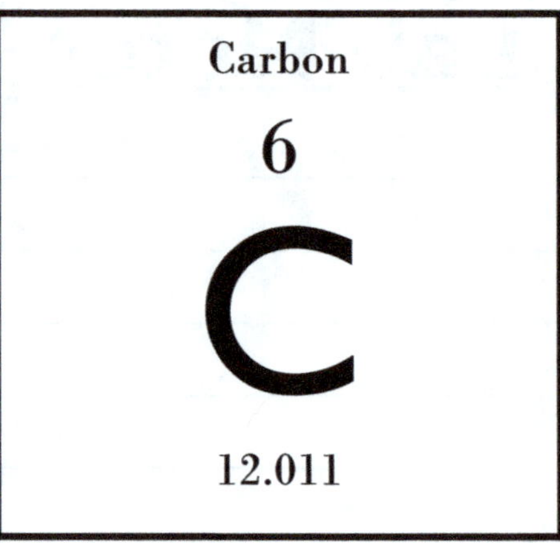
Carbon
6
C
12.011

Lab Notes

BIG IDEAS

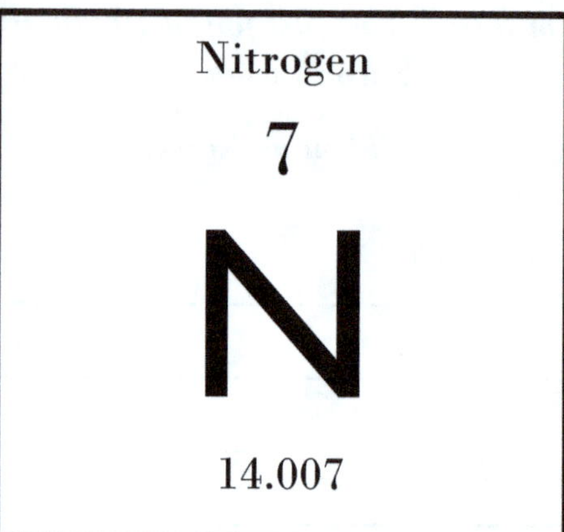

Nitrogen
7
N
14.007

Lab Notes

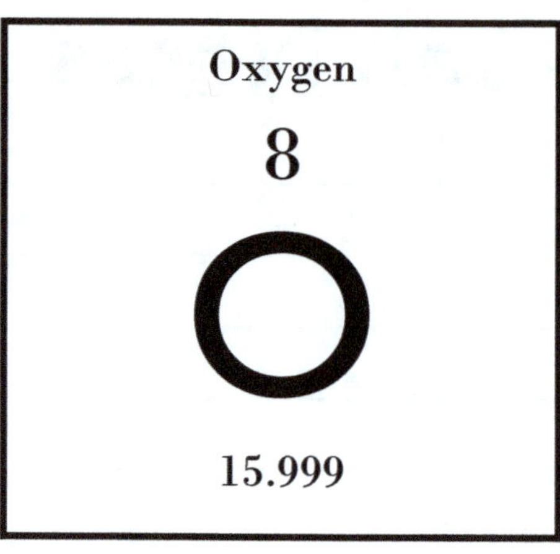

Oxygen
8
O
15.999

Lab Notes

BIG IDEAS

Fluorine
9
F
18.998

Lab Notes

Neon

10

Ne

20.180

Lab Notes

BIG IDEAS

Count what is countable, measure what is measurable, and what is not measurable, make measurable.

— *Galileo Galilei*

Sodium

11

Na

22.990

Lab Notes

Magnesium

12

Mg

24.305

Lab Notes

BIG IDEAS

Count what is countable, measure what is measurable, and what is not measurable, make measurable.

— Galileo Galilei

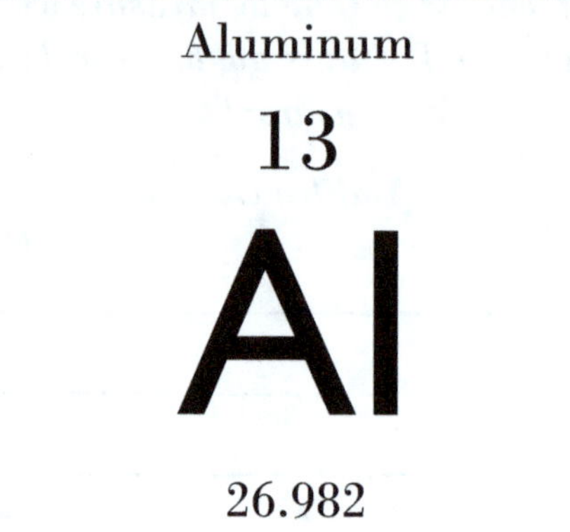

Aluminum
13
Al
26.982

Lab Notes

Silicon

14

Si

28.086

Lab Notes

BIG IDEAS

Phosphorus

15

P

30.947

Lab Notes

Sulfur

16

S

32.065

Lab Notes

BIG IDEAS

I am a firm believer, that without speculation there is no good and original observation.

—Charles Darwin

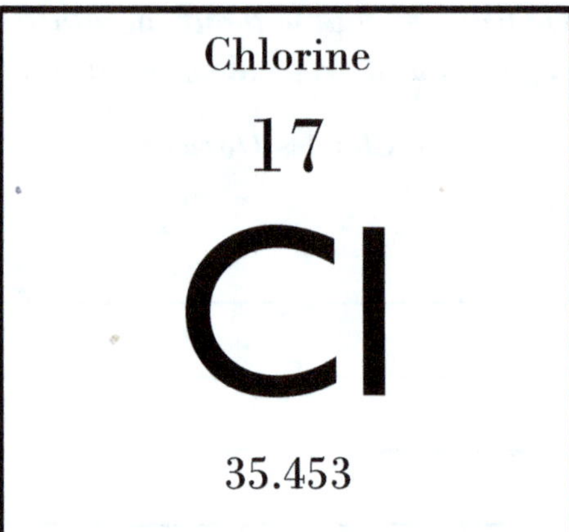

Chlorine
17
Cl
35.453

Lab Notes

Argon
18
Ar
39.948

Lab Notes

BIG IDEAS

That is the essence of science: ask an impertinent question, and you are on your way to the pertinent answer.

—Jacob Bronowski

Potassium

19

K

39.098

Lab Notes

Calcium

20

Ca

40.078

Lab Notes

BIG IDEAS

*Science is a way of thinking much more than it is a
body of knowledge.*

—Carl Sagan

Scandium

21

Sc

44.956

Lab Notes

Titanium

22

Ti

47.867

Lab Notes

BIG IDEAS

Science is the acceptance of what works and the rejection of what does not. That needs more courage than we might think.

—Jacob Bronowski

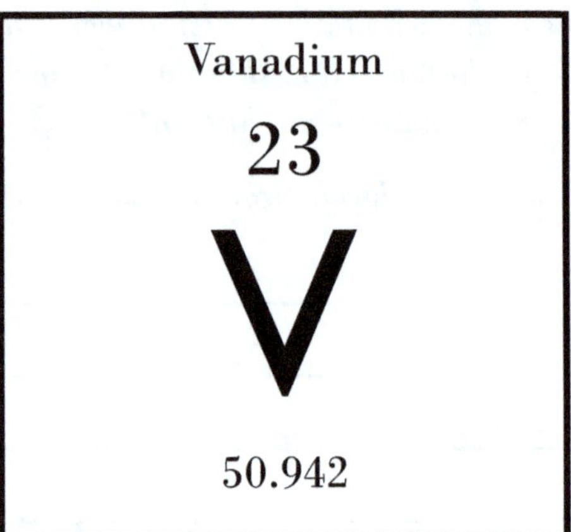

Vanadium
23
V
50.942

Lab Notes

Chromium

24

Cr

51.996

Lab Notes

BIG IDEAS

Equipped with his five senses, man explores the universe around him and calls the adventure Science.

—*Edwin Hubble*

Manganese

25

Mn

54.938

Lab Notes

Iron

26

Fe

55.845

Lab Notes

BIG IDEAS

Cobalt
27
Co
58.933

Lab Notes

Nickel
28
Ni
58.693

Lab Notes

BIG IDEAS

Progress is made by trial and failure; the failures are generally a hundred times more numerous than the successes; yet they are usually left unchronicled.

—William Ramsay

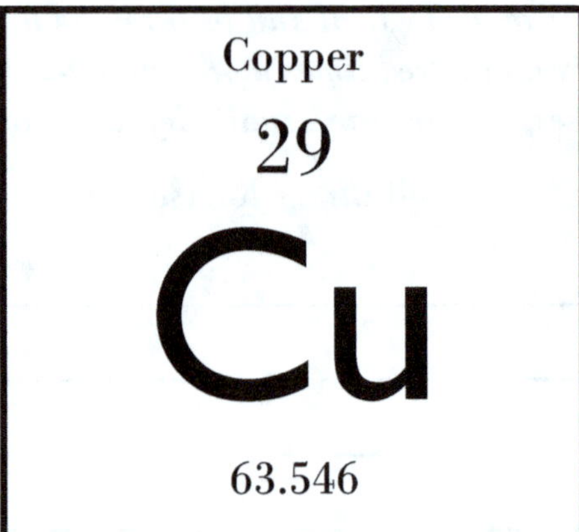

Copper
29
Cu
63.546

Lab Notes

Zinc

30

Zn

65.39

Lab Notes

BIG IDEAS

To raise new questions, new possibilities, to regard old problems from a new angle, requires creative imagination and marks real advance in science.

—Albert Einstein

Gallium

31

Ga

69.723

Lab Notes

Germanium

32

Ge

72.61

Lab Notes

BIG IDEAS

Science is not only a disciple of reason but, also, one of romance and passion.

—Stephen Hawking

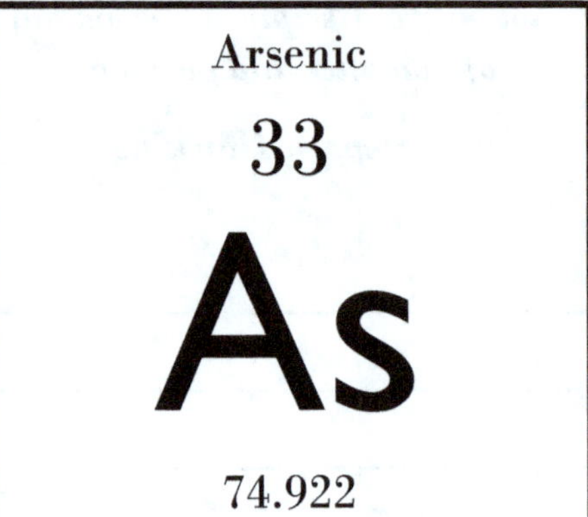

Arsenic
33
As
74.922

Lab Notes

Selenium

34

Se

78.96

Lab Notes

BIG IDEAS

Somewhere, something incredible is waiting to be known.

—Carl Sagan

Bromine

35

Br

79.904

Lab Notes

Krypton

36

Kr

83.80

Lab Notes